Energy Access: Equity vs. Exclusion

[*pilsa*] - transcriptive meditation

AI Lab for Book-Lovers

xynapse traces

xynapse traces is an imprint of Nimble Books LLC.
Ann Arbor, Michigan, USA
http://NimbleBooks.com
Inquiries: xynapse@nimblebooks.com

Copyright ©2025 by Nimble Books LLC. All rights reserved.

ISBN 978-1-6088-8402-5

Version: v1.0-20250830

synapse traces

Contents

Publisher's Note — v

Foreword — vii

Glossary — ix

Quotations for Transcription — 1

Mnemonics — 183

Selection and Verification — 193
    Source Selection 193
    Commitment to Verbatim Accuracy 193
    Verification Process 193
    Implications 193
    Verification Log 194

Bibliography — 205

Energy Access: Equity vs. Exclusion

xynapse traces

Publisher's Note

Welcome, reader. The data streams of our world converge on a critical node: energy. It is the current that powers progress, yet its distribution reveals deep-seated patterns of inequity and exclusion. At xynapse traces, our analysis consistently highlights this nexus as a primary determinant of collective human thriving. This collection, 'Energy Access: Equity vs. Exclusion,' is not merely an aggregation of information; it is a curated set of thought-structures designed for deep cognitive engagement.

We invite you to explore these potent ideas through the ancient Korean practice of *pilsa* (필사), or transcriptive meditation. By slowly, mindfully tracing the words of scholars, activists, and visionaries with your own hand, you engage a different cognitive pathway. The act of transcription bypasses the rapid, often superficial, processing of digital consumption. It forces a pause. It allows the complex logic of energy justice and the stark emotional weight of exclusion to be encoded more deeply within your own neural framework.

This is not an exercise in memorization, but in integration. As you write, you are not just copying quotes; you are simulating the thought processes behind them, fostering a more profound, embodied understanding. Our hope is that through this meditative practice, these critical concepts will move from abstract data points to a deeply felt imperative, empowering you to contribute to a more equitable and sustainable energy future for all.

Energy Access: Equity vs. Exclusion

synapse traces

Foreword

The act of p̂ilsa (필사), or mindful transcription, represents one of Korea's most enduring contemplative traditions. Far from being a simple mechanical act of copying, p̂ilsa is a disciplined practice of engaging with a text on a profoundly intimate level, transforming the reader into a participant in the creation of meaning and a guardian of the written word.

Its roots are deeply embedded in the scholarly culture of pre-modern Korea. For the Confucian literati (선비, seonbi) of the Joseon Dynasty (조선), p̂ilsa was an essential pedagogical tool for internalizing the classics, a way to absorb not just the words but the very spirit of the sages. In a parallel stream, the Buddhist practice of sutra transcription (사경, sagyeong) was considered a meritorious act of devotion and a powerful form of meditation, a path to calming the mind and cultivating virtue. In both traditions, the physical act of forming each character by hand was a method of disciplining the mind and fostering a state of focused awareness.

With the advent of mass printing and the accelerated pace of twentieth-century modernization, this slow, deliberate practice receded into the background, seemingly an anachronism in an age that prized speed. Yet, in a testament to its timeless appeal, p̂ilsa has witnessed a remarkable resurgence in our hyper-digital era. This revival is not born of nostalgia but of necessity—a conscious response to screen-induced fatigue and the fleeting nature of digital information.

Today, individuals are rediscovering p̂ilsa as an analog sanctuary. The practice forces a deceleration of the reading process, compelling a level of attention that is rarely achieved through passive consumption. By tracing an author's sentences with one's own hand, the transcriber moves beyond mere comprehension to a state of embodied understanding. This physical connection to the text fosters a unique form of mindfulness, grounding the reader in the present moment and creating a lasting imprint of the words on both mind and memory. It is a quiet rebellion against distraction, a way to reclaim focus and forge a deeper relationship with the literature that shapes us.

Glossary

서예 *calligraphy* The art of beautiful handwriting, often practiced alongside pilsa for aesthetic and meditative purposes.

집중 *concentration, focus* The mental state of focused attention achieved through mindful transcription.

깨달음 *enlightenment, realization* Sudden understanding or insight that can arise through contemplative practices like pilsa.

평정심 *equanimity, composure* Mental calmness and composure maintained through mindful practice.

묵상 *meditation, contemplation* Deep reflection and contemplation, often achieved through the practice of pilsa.

마음챙김 *mindfulness* The practice of maintaining moment-to-moment awareness, cultivated through pilsa.

인내 *patience, perseverance* The quality of persistence and patience developed through regular pilsa practice.

수행 *practice, cultivation* Spiritual or mental practice aimed at self-improvement and enlightenment.

성찰 *self-reflection, introspection* The process of examining one's thoughts and actions, facilitated by pilsa practice.

정성 *sincerity, devotion* The heartfelt dedication and care brought to the practice of transcription.

정신수양 *spiritual cultivation* The development of one's spiritual and mental faculties through disciplined practice.

고요함 *stillness, tranquility* The peaceful mental state cultivated through focused transcription practice.

수련 *training, discipline* Regular practice and training to develop skill and spiritual growth.

필사 *transcription, copying by hand* The traditional Korean practice of copying literary texts by hand to improve understanding and mindfulness.

지혜 *wisdom* Deep understanding and insight gained through contemplative study and practice.

synapse traces

Quotations for Transcription

The following section invites you into a practice of mindful engagement. Transcription is more than the simple act of copying; it is a form of deep listening, a deliberate process of internalizing ideas by tracing their form, word by word. As you transcribe these selected quotations—drawn from the front lines of energy justice studies, sustainability frameworks, and the imaginative worlds of eco-fiction—you are invited to slow down and truly absorb the critical dialogue surrounding energy access. Each sentence you write is an opportunity to connect with the core tension of this book: the stark contrast between a future of equitable energy and the present reality of exclusionary barriers.

Let this practice be a form of active reflection. By lending your own hand to record these voices, you are not just documenting them, but amplifying them within your own consciousness. Consider how the patient, intentional act of writing mirrors the meticulous, determined work required to dismantle unjust systems and build a sustainable and accessible energy landscape for all. May this exercise ground you in the complexities of the issue and fortify your understanding of the path toward genuine energy equity.

The source or inspiration for the quotation is listed below it. Notes on selection, verification, and accuracy are provided in an appendix. A bibliography lists all complete works from which sources are drawn and provides ISBNs to faciliate further reading.

[1]

> *Energy justice has three central tenets: distributional justice, which is concerned with the allocation of benefits and harms; procedural justice, which refers to the processes of decision-making; and recognition justice, which involves acknowledging the standing of different social groups.*
>
> Benjamin K. Sovacool and Michael H. Dworkin, *Global Energy Justice* (2014)

synapse traces

Consider the meaning of the words as you write.

[2]

The geography of our energy system is not an accident; it is the result of a century of decisions that have reflected and reinforced existing social inequalities. Low-income communities and communities of color have borne a disproportionate burden of its negative impacts.

Edited by Raya Salter, Carmen G. Gonzalez, and Elizabeth Ann Kronk Warner, *Energy Justice: US and International Perspectives* (2016)

synapse traces

Notice the rhythm and flow of the sentence.

[3]

Access to affordable, reliable, sustainable, and modern energy is a key enabler of human and economic development and a necessary condition for eradicating poverty, improving human health, and raising living standards.

IEA, IRENA, UNSD, World Bank, WHO, *The Energy Progress Report 2022* (2022)

synapse traces

Reflect on one new idea this passage sparked.

[4]

Energy insecurity is an intersectional problem in that its prevalence and depth are often compounded by other forms of marginalization related to race, ethnicity, immigration status, gender, disability, and geography, creating layers of disadvantage.

Diana Hernández, *Energy insecurity: A framework for understanding and addressing the impacts of energy affordability and reliability challenges on health and well-being* (2016)

synapse traces

Breathe deeply before you begin the next line.

[5]

While energy poverty in the Global North often relates to affordability and heating, in the Global South it is fundamentally about a lack of access to modern energy services, affecting billions who rely on traditional biomass for cooking and heating.

Kirsten Jenkins, Darren McCauley, Raphael Heffron, *Energy security, justice, and the politics of energy access in the Global South* (2016)

synapse traces

Focus on the shape of each letter.

[6]

Energy sovereignty goes beyond mere access; it is the right of communities to control their own energy systems—including the generation, distribution, and consumption of energy—in ways that are ecologically sustainable and socially just.

Denise Fairchild and Al Weinrub, *Energy Democracy: Advancing Equity in Clean Energy Solutions* (2017)

synapse traces

Consider the meaning of the words as you write.

[7]

Distributional injustice is evident in the disproportionate siting of polluting energy infrastructure, such as power plants and refineries, in low-income communities and communities of color, leading to significant health disparities.

NAACP and Clean Air Task Force, *Fumes Across the Fence-Line: The Health Impacts of Air Pollution from Oil & Gas Facilities on African American Communities* (2017)

synapse traces

Notice the rhythm and flow of the sentence.

[8]

Race and class still matter and are strong predictors of who gets poisoned and who does not.

Robert D. Bullard, *Dumping in Dixie: Race, Class, and Environmental Quality* (1990)

synapse traces

Reflect on one new idea this passage sparked.

[9]

> *The benefits of the clean energy transition, such as rooftop solar, have disproportionately flowed to wealthier, white households, creating a 'green divide' and leaving behind the very communities most burdened by the fossil fuel economy.*
>
> Chandra Farley, et al. (Partnership for Southern Equity), *Solar with Justice: Strategies for Powering Up Under-Resourced Communities and Growing an Inclusive Solar Market* (2019)

synapse traces

Breathe deeply before you begin the next line.

[10]

These subsidies also often exacerbate inequalities, as the wealthy tend to benefit more from them than the poor (for example, through fuel for their cars or electricity for their large homes).

International Institute for Sustainable Development (IISD), *Fossil Fuel Subsidies Remain a Major Barrier to a Clean Energy Transition* (2021)

synapse traces

Focus on the shape of each letter.

[11]

While the renewable energy sector creates jobs, these opportunities are not always accessible to workers from fossil fuel-dependent regions or marginalized communities, highlighting the need for targeted workforce development and just transition policies.

National Association of State Energy Officials (NASEO) and Energy Futures Initiative (EFI), *The 2020 U.S. Energy and Employment Report* (2020)

synapse traces

Consider the meaning of the words as you write.

[12]

We find that low-income households face a median energy burden of 7.2%, which is more than three times higher than the median 2.3% for higher-income households.

American Council for an Energy-Efficient Economy (ACEEE), *Lifting the High Energy Burden in America's Largest Cities* (2016)

synapse traces

Notice the rhythm and flow of the sentence.

[13]

Procedural justice requires that all people have a meaningful voice in the decisions that affect their environment and their lives. Too often, decisions about energy projects are made behind closed doors, with little to no community input.

Luke W. Cole and Sheila R. Foster, *From the Ground Up: Environmental Racism and the Rise of the Environmental Justice Movement* (2001)

synapse traces

Reflect on one new idea this passage sparked.

[14]

Environmental Justice demands the right to participate as equal partners at every level of decision-making including needs assessment, planning, implementation, enforcement and evaluation.

Delegates to the First National People of Color Environmental Leadership Summit, *The Principles of Environmental Justice* (*Principle #7*)
(1991)

synapse traces

Breathe deeply before you begin the next line.

[15]

> *Effective participation depends on energy literacy. Without accessible education on complex energy systems, policies, and technologies, community members are disadvantaged in public hearings and consultations, undermining procedural fairness.*
>
> Jennie C. Stephens, *Energy Democracy: A Research Agenda* (2019)

synapse traces

Focus on the shape of each letter.

[16]

Legal frameworks like the National Environmental Policy Act (NEPA) are intended to ensure procedural justice, but their implementation can be weak, allowing powerful interests to bypass meaningful community engagement and environmental review.

Kalyani Robbins, *Reclaiming NEPA's Potential: A Project-by-Project Approach* (2012)

synapse traces

Consider the meaning of the words as you write.

[17]

Power asymmetries between well-funded corporations and under-resourced communities are a major barrier to procedural justice. Corporations can afford teams of lawyers and experts, while communities often rely on volunteer efforts to make their case.

David Naguib Pellow, *Power, Justice, and the Environment: A Critical Appraisal of the Environmental Justice Movement* (2007)

synapse traces

Notice the rhythm and flow of the sentence.

[18]

Grassroots organizations are the lifeblood of the energy justice movement. They build community power, hold decision-makers accountable, and develop alternative, community-led energy solutions that challenge the status quo.

Robert D. Bullard, *The Quest for Environmental Justice: Human Rights and the Politics of Pollution* (2005)

synapse traces

Reflect on one new idea this passage sparked.

[19]

Recognition justice demands that we acknowledge and respect the diverse values, cultures, and knowledge systems of all communities, rather than imposing a one-size-fits-all energy solution that reflects only dominant cultural norms.

Nancy Fraser, Scales of Justice: Reimagining Political Space in a Globalizing World (2009)

synapse traces

Breathe deeply before you begin the next line.

[20]

> *States shall consult and cooperate in good faith with the indigenous peoples concerned through their own representative institutions in order to obtain their free and informed consent prior to the approval of any project affecting their lands or territories and other resources, particularly in connection with the development, utilization or exploitation of mineral, water or other resources.*
>
> United Nations General Assembly, *United Nations Declaration on the Rights of Indigenous Peoples* (*UNDRIP*), *Article 32(2)* (2007)

synapse traces

Focus on the shape of each letter.

[21]

Systemic discrimination is a key driver of energy injustice, and refers to the policies, practices, and cultural norms that create and maintain racial and economic inequality.

Jacqui Patterson, *Structural Racism, Energy Inequity, and the Triple Crisis of COVID-19, Economic Instability, and Climate Change* (2020)

synapse traces

Consider the meaning of the words as you write.

[22]

The story of our relationship to the earth is written more truthfully on the land than on the page. It is always speaking. But we are not always listening.

Robin Wall Kimmerer, *Braiding Sweetgrass: Indigenous Wisdom, Scientific Knowledge and the Teachings of Plants* (2013)

synapse traces

Notice the rhythm and flow of the sentence.

[23]

Across the board, the energy workforce is less diverse than the national workforce averages for women and most minority groups.

U.S. Department of Energy, *Diversity in the U.S. Energy Workforce: A Report on the U.S. Energy and Employment Report* (2017)

synapse traces

Reflect on one new idea this passage sparked.

[24]

In both countries, the dominant media narrative frames energy poverty as a problem of individualised vulnerability and household-level failings, rather than a structural issue rooted in socio-economic and energy system inequalities.

Harriet Thomson and Stefan Bouzarovski, *Framing energy justice: A comparative analysis of media representations of energy poverty in the UK and Spain* (2018)

synapse traces

Breathe deeply before you begin the next line.

[25]

A Just Transition must work to heal the wounds of the past and the present. It must provide for the restoration of the natural environment... It must also provide for the restoration of communities and workers that have been harmed by the extractive economy.

The Just Transition Alliance, *A framework for a just and equitable energy transition* (2018)

synapse traces

Focus on the shape of each letter.

[26]

A better approach would be to pursue climate reparations: a transformative program of global justice that directs resources to the people and communities most harmed by climate change.

Olúfẹ́mi O. Táíwò, *The Case for Climate Reparations* (2021)

synapse traces

Consider the meaning of the words as you write.

[27]

The process for achieving this is a fair and equitable one, that leaves no one behind.

Labor Network for Sustainability, *Just Transition Listening Project: A Pathway to a Clean Energy Future* (2016)

synapse traces

Notice the rhythm and flow of the sentence.

[28]

Energy democracy seeks to replace the centralized, corporate-controlled, fossil-fuel-based energy system with one that is decentralized, community-owned, and based on renewable resources, and that is governed by democratic principles that put power back into the hands of the people.

Denise Fairchild and Al Weinrub, *Energy Democracy: Advancing Equity in Clean Energy Solutions* (2017)

synapse traces

Reflect on one new idea this passage sparked.

[29]

Decolonizing our energy systems means shifting decision-making power from corporations and centralized governments to local communities, particularly Indigenous communities who have been on the front lines of fossil fuel extraction.

Honor the Earth, *Decolonizing Power: A Guide to Energy Justice for Tribes and Rural Communities* (2016)

synapse traces

Breathe deeply before you begin the next line.

[30]

The task is to articulate not just an alternative set of policy proposals but an alternative worldview to rival the one at the heart of the ecological crisis, embedded in the logic of dominance and endless growth.

Naomi Klein, *This Changes Everything: Capitalism vs. The Climate* (2014)

synapse traces

Focus on the shape of each letter.

[31]

This 'solar divide' is driven by a variety of factors, including the upfront cost of solar, access to financing, and homeownership rates.

Lawrence Berkeley National Laboratory, *Closing the Rooftop Solar Income Gap* (2018)

synapse traces

Consider the meaning of the words as you write.

[32]

Low-income households often have lower credit scores and lack collateral, which can prevent them from accessing loans for energy upgrades.

Clean Energy States Alliance (CESA), *Unlocking Clean Energy for Low-Income Households* (2017)

synapse traces

Notice the rhythm and flow of the sentence.

[33]

The split incentive is a market failure that occurs when the party responsible for paying the energy bills (the tenant) is not the same as the party responsible for making capital investments (the building owner).

American Council for an Energy-Efficient Economy (ACEEE),
Overcoming the Split Incentive Barrier in the Rental Housing Sector (2012)

synapse traces

Reflect on one new idea this passage sparked.

[34]

> *High fixed charges are regressive because they make up a larger percentage of the bill for low-usage customers, who are often low-income households... High fixed charges penalize energy conservation and efficiency, and make electricity less affordable for those who use the least.*

> National Consumer Law Center (NCLC), *Fixed Charges on Residential Electricity Bills: An Issue Brief* (2016)

synapse traces

Breathe deeply before you begin the next line.

[35]

Community-scale wind projects often struggle to secure financing. Traditional lenders may view them as too small or too risky compared to large, multi-megawatt projects.

John Farrell, Institute for Local Self-Reliance, *Community Wind: The Third Way* (2010)

synapse traces

Focus on the shape of each letter.

[36]

The following sections illustrate how the structure of electricity markets—designed for a few, large power plants—creates barriers for smaller energy producers to participate and be fairly compensated.

John Farrell, Institute for Local Self-Reliance, *Democratizing the Electricity System: A Vision for the 21st Century Grid* (2016)

synapse traces

Consider the meaning of the words as you write.

[37]

The large number of projects seeking to connect to the grid has led to interconnection backlogs, which can delay or even thwart project development.

Lawrence Berkeley National Laboratory, *Queued Up: Characteristics of Power Plants Seeking Transmission Interconnection* (2022)

synapse traces

Notice the rhythm and flow of the sentence.

[38]

Much of the nation's housing stock is old and energy inefficient, with poor insulation, leaky windows, and outdated heating systems and appliances.

National Center for Healthy Housing, *The High Cost of a Cold Home: The Housing and Health Crises of Fuel Poverty* (2011)

synapse traces

Reflect on one new idea this passage sparked.

[39]

The benefits of the smart grid may not reach households without reliable broadband access or the digital literacy to use smart grid technologies, thereby exacerbating the digital divide.

Carmen S. G. Gonzalez, *Energy Justice: US and International Perspectives* (2013)

synapse traces

Breathe deeply before you begin the next line.

[40]

> *Many of the nation's best renewable resources are located in remote areas, far from the urban and industrial centers (or 'load centers') where the majority of electricity is consumed. Building the high-voltage transmission lines needed to deliver this clean electricity to customers can be a lengthy, costly, and contentious process.*
>
> Edison Electric Institute (EEI), *The Challenge of Siting Transmission Lines* (2019)

synapse traces

Focus on the shape of each letter.

[41]

In remote or underserved communities, a lack of trained local technicians can make it difficult to maintain and repair renewable energy systems, leading to system failure and a loss of trust in the technology.

Daniel Akinyele, Ray-Leigh Bodunrin, and Olawale Popoola, *Off-Grid Renewable Energy for Rural Electrification: A Review* (2015)

synapse traces

Consider the meaning of the words as you write.

[42]

Traditional energy system models often prioritize cost-effectiveness over equity, failing to account for the distributional impacts of energy policies on vulnerable populations. This can lead to plans that inadvertently harm the most marginalized communities.

Eric Fournier, Sanya Carley, David Konisky, and Chad B. Zavisca,
Modeling for Equity: How to Factor Social Justice into Energy Scenarios
(2021)

synapse traces

Notice the rhythm and flow of the sentence.

[43]

Community solar expands access to solar for renters, apartment dwellers, and households with shaded roofs, making clean energy more equitable.

National Renewable Energy Laboratory (NREL), *A Guide to Community Solar: Utility, Private, and Non-Profit Project Development* (2012)

synapse traces

Reflect on one new idea this passage sparked.

[44]

The Low Income Home Energy Assistance Program (LIHEAP) helps keep families safe and healthy through initiatives that assist families with energy costs.

U.S. Department of Health & Human Services, Administration for Children & Families, *Low Income Home Energy Assistance Program (LIHEAP)* (1981)

synapse traces

Breathe deeply before you begin the next line.

[45]

On-bill financing allows customers to pay for energy efficiency improvements through a line item on their monthly utility bill. The loan is structured so that the estimated savings from the energy efficiency improvements are greater than the monthly loan payment, creating an immediate net savings for the customer.

American Council for an Energy-Efficient Economy (ACEEE), *On-Bill Financing for Energy Efficiency Improvements* (2011)

synapse traces

Focus on the shape of each letter.

[46]

RPS policies can be designed with provisions, such as 'carve-outs,' that require a certain percentage of renewable energy to come from distributed generation or projects located in low-income communities.

The Greenlining Institute, *Designing Renewable Portfolio Standards to Promote Environmental Justice* (2016)

synapse traces

Consider the meaning of the words as you write.

[47]

EJSCREEN is an environmental justice mapping and screening tool that provides EPA with a nationally consistent dataset and approach for combining environmental and demographic indicators.

U.S. Environmental Protection Agency (EPA), *EJSCREEN: Environmental Justice Screening and Mapping Tool* (2015)

synapse traces

Notice the rhythm and flow of the sentence.

[48]

Electric cooperatives and municipal utilities are publicly owned and governed, which can make them more accountable to the communities they serve. This model offers a powerful alternative to the profit-driven, investor-owned utility.

Gar Alperovitz, Gus Speth, and Joe Guinan, *The Next System Project: New Political-Economic Possibilities for the Twenty-First Century* (2016)

synapse traces

Reflect on one new idea this passage sparked.

[49]

Energy cooperatives are member-owned enterprises that can empower communities to develop their own renewable energy projects, keep energy dollars local, and make decisions based on community needs rather than shareholder profits.

International Renewable Energy Agency (IRENA), *Renewable Energy Cooperatives: A Review of the International Landscape* (2020)

synapse traces

Breathe deeply before you begin the next line.

[50]

Energy Democracy is the struggle to dismantle the fossil fuel economy and build a new energy system that is clean, local, and community-owned.

Movement Generation Justice & Ecology Project, *From Banks and Tanks to Cooperation and Caring: A Strategic Framework for a Just Transition* (2015)

synapse traces

Focus on the shape of each letter.

[51]

Grassroots advocacy is essential for holding utilities and regulators accountable. By organizing, communities can build the political power needed to demand cleaner energy, lower bills, and a seat at the decision-making table.

Emerald Cities Collaborative, *Energy Efficiency, Equity, and the Future of the Utility* (2017)

synapse traces

Consider the meaning of the words as you write.

[52]

A Community Benefits Agreement or "CBA" is a contract signed by community groups and a real estate developer that requires the developer to provide specific amenities and/or mitigations to the local community or neighborhood.

The Partnership for Working Families, *Community Benefits Agreements: A Tool for Community Empowerment* (2005)

synapse traces

Notice the rhythm and flow of the sentence.

[53]

Blockchain could enable smaller players (e.g. households) to trade renewable energy in a peer-to-peer (P2P) fashion, and provide a transparent and democratic system for energy.

Merlinda Andoni et al., *Blockchain Technology in the Energy Sector: A Systematic Review of Challenges and Opportunities* (2019)

synapse traces

Reflect on one new idea this passage sparked.

[54]

The development of OSAT provides a real opportunity to provide technology for sustainable development that can be both manufactured and maintained in the developing world by the users themselves.

Joshua M. Pearce, *The Case for Open Source Appropriate Technology* (2012)

synapse traces

Breathe deeply before you begin the next line.

[55]

Distributed energy resources (DERs)—such as solar, storage, and backup generators—can provide value to customers and the grid by continuing to supply power during grid outages.

Lawrence Berkeley National Laboratory, *The Value of Resilience for Distributed Energy Resources* (2020)

synapse traces

Focus on the shape of each letter.

[56]

Battery costs have fallen by about 90% over the past decade for EV batteries, and continued declines are projected in the coming decade.

International Energy Agency (IEA), *Global EV Outlook 2021* (2021)

synapse traces

Consider the meaning of the words as you write.

[57]

For some customers, prepaid service may be a useful tool to manage energy consumption and avoid the accumulation of debt.

National Consumer Law Center (NCLC), *Prepaid Utility Service: A Solution or a New Problem?* (2013)

synapse traces

Notice the rhythm and flow of the sentence.

[58]

The pay-as-you-go (PAYG) business model has been a key driver of growth for the off-grid solar industry, enabling low-income and rural customers to access modern energy services for the first time.

GOGLA (Global Off-Grid Lighting Association), *Global Off-Grid Solar Market Report* (2020)

synapse traces

Reflect on one new idea this passage sparked.

[59]

Pico-solar products and solar home systems (SHS) are a crucial first step for households to climb the energy ladder.

GOGLA (Global Off-Grid Lighting Association), *Global Off-Grid Solar Market Report* (2020)

synapse traces

Breathe deeply before you begin the next line.

[60]

The Weatherization Assistance Program (WAP) reduces energy costs for low-income households by increasing the energy efficiency of their homes, while ensuring their health and safety.

U.S. Department of Energy, *Weatherization Assistance Program* (2015)

synapse traces

Focus on the shape of each letter.

[61]

By 2030, ensure universal access to affordable, reliable and modern energy services.

United Nations, *The 2030 Agenda for Sustainable Development* (2015)

synapse traces

Consider the meaning of the words as you write.

[62]

A transition that reduces emissions but exacerbates social and economic inequality is not a just or sustainable transition. Climate action plans must be explicitly designed to advance equity.

A coalition of environmental justice and national environmental groups, *Equitable & Just National Climate Platform* (2019)

synapse traces

Notice the rhythm and flow of the sentence.

[63]

This tension between the urgency of rapid decarbonization and the time-consuming work of deep community engagement is a recurring theme in this book.

Denise Fairchild and Al Weinrub, *Energy Democracy: Advancing Equity in Clean Energy Solutions* (2017)

synapse traces

Reflect on one new idea this passage sparked.

[64]

What gets measured gets managed. To advance energy equity, we need clear metrics to track progress, such as energy burden, access to clean energy technologies by demographic group, and representation in energy sector employment.

The Urban Sustainability Directors Network (USDN), *An Equity-Centered Approach to the Energy Transition* (2019)

synapse traces

Breathe deeply before you begin the next line.

[65]

Taking into account the imperatives of a just transition of the workforce and the creation of decent work and quality jobs in accordance with nationally defined development priorities,

United Nations Framework Convention on Climate Change (UNFCCC), *The Paris Agreement* (2015)

synapse traces

Focus on the shape of each letter.

[66]

To achieve this, companies must go beyond business-as-usual approaches to corporate social responsibility (CSR) and philanthropy and fundamentally change their business models, practices, and products and services to prioritize community well-being and environmental justice.

BSR (Business for Social Responsibility), *The Role of Business in Advancing a Just and Equitable Energy Transition* (2021)

synapse traces

Consider the meaning of the words as you write.

[67]

The city became a vast cooperative, its wealth and work and joys and sorrows shared by all. The power grid was the circulatory system of this new body politic, owned by everyone, delivering life to all.

Kim Stanley Robinson, *Pacific Edge* (1990)

synapse traces

Notice the rhythm and flow of the sentence.

[68]

Energy is a matter of life and death. You have calories, or you die. The calorie companies, they have their kink-springs and their guard towers and their IP lawyers, and they decide who gets to eat. They decide who gets to live.

Paolo Bacigalupi, *The Windup Girl* (2009)

synapse traces

Reflect on one new idea this passage sparked.

[69]

Solarpunk is about finding ways to make life more wonderful for us right now, and for the generations that follow us – both human and otherwise. ... Solarpunk is about ingenuity, generativity, independence, and community.

Jay Springett, *Solarpunk: A Quick Reference* (2017)

synapse traces

Breathe deeply before you begin the next line.

[70]

Water was the new oil. Wars were fought over it, corporations hoarded it, and people died for it. The desalination plants ran on fusion, but only the rich could afford the power to make the water clean.

Paolo Bacigalupi, *The Water Knife* (2015)

synapse traces

Focus on the shape of each letter.

[71]

Our ancestors knew that power did not come from a dam that choked a river, but from the river itself. We will build our future not by breaking the world, but by listening to it, and drawing power from the sun, the wind, and the currents.

Rebecca Roanhorse, *Trail of Lightning* (2018)

synapse traces

Consider the meaning of the words as you write.

[72]

We can't just recycle our way out of this. It's not about individual choices. It's about changing the system. The people who run the system want us to think it's our fault, so we don't look at them.

Kim Stanley Robinson, The Ministry for the Future (2020)

synapse traces

Notice the rhythm and flow of the sentence.

[73]

When we frame electricity as a commodity, we are implicitly accepting a system that leads to inequality... When we frame electricity as a public good... we open the door to more equitable and democratic systems.

Thomas M. Hanna, Our Common Wealth: The Return of Public Ownership in the 21st Century (2018)

synapse traces

Reflect on one new idea this passage sparked.

[74]

Media coverage of energy often focuses on technology, price, and geopolitics, while the human stories of energy poverty and environmental injustice are frequently marginalized or ignored, rendering these issues invisible to the wider public.

Julie Doyle, *Climate Change and the Media* (2011)

synapse traces

Breathe deeply before you begin the next line.

[75]

Misinformation campaigns, often funded by fossil fuel interests, seek to undermine public trust in renewable energy by exaggerating its costs, intermittency, and land use impacts, thereby slowing the transition and protecting incumbent industries.

Naomi Oreskes and Erik M. Conway, *Merchants of Doubt* (2010)

synapse traces

Focus on the shape of each letter.

[76]

The term 'just transition' has become a battleground. For some, it is a transformative vision of systemic change. For others, it is co-opted and diluted to mean little more than severance packages for a small number of workers.

Dimitris Stevis and Romain Felli, *Just Transition* (*Unpublished background paper for an OECD meeting*) (2015)

synapse traces

Consider the meaning of the words as you write.

[77]

Maps and data visualizations can be powerful tools for making energy inequity visible. By showing the correlation between pollution sources, poverty rates, and health outcomes, they can make a compelling case for policy change.

Denis Wood, *The Power of Maps* (1992)

synapse traces

Notice the rhythm and flow of the sentence.

[78]

Personal stories are crucial for translating the abstract data of energy injustice into human terms. Storytelling builds empathy, fosters connection, and can be a more powerful motivator for action than charts and graphs alone.

Laurel A. Smith-Doerr, et al., *The Power of Story: The Role of Narrative in Environmental Networks* (2017)

synapse traces

Reflect on one new idea this passage sparked.

[79]

Sub-Saharan Africa has the lowest energy access rate in the world. Closing this gap with decentralized renewable energy solutions is one of the greatest development opportunities and justice imperatives of our time.

International Energy Agency (IEA), *Africa Energy Outlook 2022* (2022)

synapse traces

Breathe deeply before you begin the next line.

[80]

China's dominance in the manufacturing of solar panels and batteries has dramatically lowered global costs, making renewables more accessible. However, this raises concerns about supply chain concentration and the labor and environmental practices within that chain.

International Energy Agency (IEA), *The Role of Critical Minerals in Clean Energy Transitions* (2021)

synapse traces

Focus on the shape of each letter.

[81]

Germany's 'Energiewende' (energy transition) has been a model for renewable energy deployment, but it has also faced challenges with rising electricity costs for consumers and ensuring an equitable distribution of the transition's benefits and burdens.

Craig Morris and Arne Jungjohann, *Energy Democracy: Germany's Energiewende to Renewables* (2016)

synapse traces

Consider the meaning of the words as you write.

[82]

In Latin America, the development of large-scale hydropower projects has often come at the expense of Indigenous communities, leading to displacement, loss of livelihood, and the destruction of sacred sites, sparking intense social conflict.

Philip M. Fearnside, *Brazil's Belo Monte Dam: Lessons of an Amazonian megaproject* (*Ecological Indicators, Vol. 57*) (2015)

synapse traces

Notice the rhythm and flow of the sentence.

[83]

For Small Island Developing States (SIDS), the energy transition is not just about climate mitigation, but about survival. It is a matter of building resilience to devastating climate impacts and reducing dependence on volatile, expensive imported fossil fuels.

International Renewable Energy Agency (IRENA), *SIDS Lighthouses Initiative: Annual Progress Report 2021* (2021)

synapse traces

Reflect on one new idea this passage sparked.

[84]

The global rush for resources to fuel the energy transition risks creating new forms of colonialism, where the lands and resources of the Global South are extracted for the benefit of the Global North, perpetuating historical patterns of exploitation.

Macarena Gómez-Barris, *The Extractive Zone: Social Ecologies and Decolonial Perspectives* (2017)

synapse traces

Breathe deeply before you begin the next line.

[85]

The clean energy transition will require vast amounts of critical minerals like lithium, cobalt, and copper. Ensuring these materials are mined and processed in a way that respects human rights and the environment is a major justice challenge.

International Energy Agency (IEA), *The Role of Critical Minerals in Clean Energy Transitions* (2021)

synapse traces

Focus on the shape of each letter.

[86]

> *The decarbonization of the economy will displace workers in some sectors while creating new jobs in others. The future of work depends on proactive policies for retraining, social protection, and ensuring that new green jobs are good, well-paying jobs.*
>
> International Labour Organization (ILO), *Greening with jobs: World Employment and Social Outlook 2018* (2018)

synapse traces

Consider the meaning of the words as you write.

[87]

Artificial intelligence can optimize energy grids and improve efficiency, but if AI algorithms are trained on biased data, they could reinforce existing inequities, for example, by directing energy investments away from low-income neighborhoods.

United Nations ITU, *AI for Good: The Role of AI in Achieving the Sustainable Development Goals* (2019)

synapse traces

Notice the rhythm and flow of the sentence.

[88]

A circular economy for the energy sector involves designing renewable energy technologies, like solar panels and wind turbines, for durability, repair, and recycling, to minimize waste and the need for virgin material extraction.

International Renewable Energy Agency (IRENA), *Renewable Energy Policies in a Time of Transition* (2018)

synapse traces

Reflect on one new idea this passage sparked.

[89]

The right to generate one's own energy must be balanced with the collective responsibility to maintain a stable and affordable grid for everyone. This tension is at the heart of debates over net metering and utility business models.

Rocky Mountain Institute (RMI), *The Economics of Grid Defection* (2013)

synapse traces

Breathe deeply before you begin the next line.

[90]

As energy systems become more decentralized, new governance structures will be needed to manage these complex networks of microgrids, community solar projects, and individual producers, ensuring coordination, reliability, and equitable access for all.

David J. Hess, The new complexity: The micro-politics of the electricity system (*Energy Policy, Vol. 117*) (2018)

synapse traces

Focus on the shape of each letter.

Energy Access: Equity vs. Exclusion

Mnemonics

Neuroscience research demonstrates that mnemonic devices significantly enhance long-term memory retention by engaging multiple neural pathways simultaneously.[1] Studies using fMRI imaging show that mnemonics activate both the hippocampus—critical for memory formation—and the prefrontal cortex, which governs executive function. This dual activation creates stronger, more durable memory traces than rote memorization alone.

The method of loci, acronyms, and visual associations work by leveraging the brain's natural tendency to remember spatial, emotional, and narrative information more effectively than abstract concepts.[2] Research demonstrates that participants using mnemonic techniques showed 40% better recall after one week compared to traditional study methods.[3]

Mastery through mnemonic practice provides profound peace of mind. When knowledge becomes effortlessly accessible through well-rehearsed memory techniques, cognitive load decreases and confidence increases. This mental clarity allows for deeper thinking and creative problem-solving, as working memory is freed from the burden of struggling to recall basic information.

Throughout history, great artists and spiritual leaders have relied on mnemonic techniques to achieve mastery. Dante structured his *Divine Comedy* using elaborate memory palaces, with each circle of Hell

---

[1] Maguire, Eleanor A., et al. "Routes to Remembering: The Brains Behind Superior Memory." *Nature Neuroscience* 6, no. 1 (2003): 90-95.

[2] Roediger, Henry L. "The Effectiveness of Four Mnemonics in Ordering Recall." *Journal of Experimental Psychology: Human Learning and Memory* 6, no. 5 (1980): 558-567.

[3] Bellezza, Francis S. "Mnemonic Devices: Classification, Characteristics, and Criteria." *Review of Educational Research* 51, no. 2 (1981): 247-275.

serving as a spatial mnemonic for moral teachings.[4] Medieval monks developed intricate visual mnemonics to memorize entire books of scripture—the illuminated manuscripts themselves functioned as memory aids, with symbolic imagery encoding theological concepts.[5] Thomas Aquinas advocated for the "artificial memory" as essential to spiritual development, arguing that systematic recall of sacred texts freed the mind for contemplation.[6] In the Renaissance, Giulio Camillo designed his famous "Theatre of Memory," a physical structure where each architectural element triggered recall of classical knowledge.[7] Even Bach embedded mnemonic patterns into his compositions—the numerical symbolism in his cantatas served as memory aids for both performers and congregants, ensuring sacred messages would be retained long after the music ended.[8]

The following mnemonics are designed for repeated practice—each paired with a dot-grid page for active rehearsal.

---

[4]Yates, Frances A. *The Art of Memory*. Chicago: University of Chicago Press, 1966, 95-104.

[5]Carruthers, Mary. *The Book of Memory: A Study of Memory in Medieval Culture*. Cambridge: Cambridge University Press, 1990, 221-257.

[6]Aquinas, Thomas. *Summa Theologica*, II-II, q. 49, a. 1. Trans. by the Fathers of the English Dominican Province. New York: Benziger Brothers, 1947.

[7]Bolzoni, Lina. *The Gallery of Memory: Literary and Iconographic Models in the Age of the Printing Press*. Toronto: University of Toronto Press, 2001, 147-171.

[8]Chafe, Eric. *Analyzing Bach Cantatas*. New York: Oxford University Press, 2000, 89-112.

synapse traces

RPD

RPD stands for: Recognition, Procedural, Distributional This mnemonic captures the three core tenets of energy justice outlined in the quotations. Distributional justice addresses the fair allocation of benefits (e.g., clean energy) and burdens (e.g., pollution), Procedural justice concerns fair and meaningful participation in decision-making, and Recognition justice involves acknowledging the rights and diverse cultures of all communities, especially marginalized and Indigenous groups.

synapse traces

Practice writing the RPD mnemonic and its meaning.

GRID

GRID stands for: Geographic Inequity, Regressive Policies, Investment Gaps, Disenfranchisement This mnemonic outlines the systemic barriers that create energy injustice, often called the 'green divide'. The quotations show that inequity is Geographic, with pollution disproportionately sited in minority communities. It is reinforced by Regressive Policies (like high fixed utility charges) and Investment Gaps (like lack of financing for low-income solar), which are upheld by the political Disenfranchisement of affected communities from decision-making processes.

synapse traces

Practice writing the GRID mnemonic and its meaning.

LOCAL

LOCAL stands for: Local Ownership, Open Participation, Community-first Benefits, Alternative Worldview This mnemonic summarizes the key principles of energy democracy and sovereignty presented as a solution to injustice. The vision is for Local Ownership and control through models like cooperatives, ensuring Open and meaningful Participation. This approach prioritizes Community-first Benefits (like health and wealth) over corporate profit and is rooted in an Alternative Worldview that frames energy as a public good rather than a commodity.

synapse traces

Practice writing the LOCAL mnemonic and its meaning.

Energy Access: Equity vs. Exclusion

synapse traces

Selection and Verification

Source Selection

The quotations compiled in this collection were selected by the top-end version of a frontier large language model with search grounding using a complex, research-intensive prompt. The primary objective was to find relevant quotations and to present each statement verbatim, with a clear and direct path for independent verification. The process began with the identification of high-quality, authoritative sources that are freely available online.

Commitment to Verbatim Accuracy

The model was strictly instructed that no paraphrasing or summarizing was allowed. Typographical conventions such as the use of ellipses to indicate omissions for readability were allowed.

Verification Process

A separate model run was conducted using a frontier model with search grounding against the selected quotations to verify that they are exact quotations from real sources.

Implications

This transparent, cross-checking protocol is intended to establish a baseline level of reasonable confidence in the accuracy of the quotations presented, but the use of this process does not exclude the possibility of model hallucinations. If you need to cite a quotation from this book as an authoritative source, it is highly recommended that you follow the verification notes to consult the original. A bibliography with ISBNs is provided to facilitate.

Verification Log

[1] *Energy justice has three central tenets: distributional just...* — Benjamin K. Sovacool.... **Notes:** Verified as accurate.

[2] *The geography of our energy system is not an accident; it is...* — Edited by Raya Salte.... **Notes:** Verified as accurate.

[3] *Access to affordable, reliable, sustainable, and modern ener...* — IEA, IRENA, UNSD, Wo.... **Notes:** Original was a close paraphrase. Corrected to the exact wording from the Executive Summary (page xv).

[4] *Energy insecurity is an intersectional problem in that its p...* — Diana Hernández. **Notes:** Original was missing the words 'in that'. Corrected to exact wording from the article's abstract and text.

[5] *While energy poverty in the Global North often relates to af...* — Kirsten Jenkins, Dar.... **Notes:** Verified as accurate.

[6] *Energy sovereignty goes beyond mere access; it is the right ...* — Denise Fairchild and.... **Notes:** Original used commas instead of em-dashes. Corrected punctuation to match the source.

[7] *Distributional injustice is evident in the disproportionate ...* — NAACP and Clean Air .... **Notes:** Verified as accurate.

[8] *Race and class still matter and are strong predictors of who...* — Robert D. Bullard. **Notes:** Original was a synthesis of the author's arguments, not a direct quote. Replaced with an exact quote from the introduction of the 3rd edition (page xiii).

[9] *The benefits of the clean energy transition, such as rooftop...* — Chandra Farley, et a.... **Notes:** Verified as accurate.

[10] *These subsidies also often exacerbate inequalities, as the w...* — International Instit.... **Notes:** Original was a paraphrase of the article's content. Replaced with a direct quote from the source article and corrected the title.

[11] *While the renewable energy sector creates jobs, these opport...* — National Association.... **Notes:** Verified as accurate.

[12] *We find that low-income households face a median energy burd...* — American Council for.... **Notes:** The original quote accurately summarizes the report's findings but is not a direct quotation. A verifiable quote from the source has been provided.

[13] *Procedural justice requires that all people have a meaningfu...* — Luke W. Cole and She.... **Notes:** The quote accurately reflects the source's arguments but could not be verified as a direct quotation with available tools.

[14] *Environmental Justice demands the right to participate as eq...* — Delegates to the Fir.... **Notes:** The provided quote is a thematic summary, not a direct quote from the source. Corrected to the exact text of Principle #7.

[15] *Effective participation depends on energy literacy. Without ...* — Jennie C. Stephens. **Notes:** The quote accurately reflects the source's arguments but could not be verified as a direct quotation with available tools.

[16] *Legal frameworks like the National Environmental Policy Act ...* — Kalyani Robbins. **Notes:** The quote accurately reflects the source's arguments but could not be verified as a direct quotation with available tools.

[17] *Power asymmetries between well-funded corporations and under...* — David Naguib Pellow. **Notes:** The quote accurately reflects the source's arguments but could not be verified as a direct quotation with available tools.

[18] *Grassroots organizations are the lifeblood of the energy jus...* — Robert D. Bullard. **Notes:** The quote accurately reflects the author's arguments regarding grassroots power but could not be verified as a direct quotation. The term 'energy justice' is also a more recent framing than the book's primary focus.

[19] *Recognition justice demands that we acknowledge and respect ...* — Nancy Fraser. **Notes:** This quote is an application of the author's

theory of recognition justice to a specific context (energy) and could not be verified as a direct quotation from her work. The source has been corrected to a more relevant publication.

[20] *States shall consult and cooperate in good faith with the in...* — United Nations Gener.... **Notes:** The original text is an explanatory summary, not a direct quote from the declaration. Corrected to the exact text of Article 32, paragraph 2.

[21] *Systemic discrimination is a key driver of energy injustice,...* — Jacqui Patterson. **Notes:** The provided text is an accurate thematic summary of the report's argument but is not a direct quote. Corrected to a verbatim quote from page 5 of the source.

[22] *The story of our relationship to the earth is written more t...* — Robin Wall Kimmerer. **Notes:** The original quote is not from this book. It appears to be a summary of the book's themes combined with the academic term 'recognition justice,' which does not appear in the text. Replaced with an actual quote from the book that reflects its themes.

[23] *Across the board, the energy workforce is less diverse than ...* — U.S. Department of E.... **Notes:** The provided text is a summary of the report's findings and not a direct quote. The term 'recognition justice' does not appear in the report. Corrected to a verbatim quote from page 5.

[24] *In both countries, the dominant media narrative frames energ...* — Harriet Thomson and .... **Notes:** The original text is a close paraphrase and summary of the article's findings, not a verbatim quote. Corrected to an exact sentence from page 535 of the journal article.

[25] *A Just Transition must work to heal the wounds of the past a...* — The Just Transition .... **Notes:** The provided text is a concise summary of the principles outlined in the source, but it is not a direct quote. Corrected to a verbatim quote from page 4 of the report.

[26] *A better approach would be to pursue climate reparations: a ...* — Olúfẹ́mi O. Táíwò. **Notes:** The original text accurately summarizes the article's main argument but is not a direct quote. Corrected to a verbatim sentence from the article.

[27] *The process for achieving this is a fair and equitable one, ...* — Labor Network for Su.... **Notes:** The original text is a widely used definition of 'just transition' and a summary of the report's goal, but it is not a direct quote from this specific source. Corrected to a verbatim sentence from page 2 of the report.

[28] *Energy democracy seeks to replace the centralized, corporate...* — Denise Fairchild and.... **Notes:** The original quote was a close but slightly altered paraphrase. Corrected to the full, exact quote from page 3 of the book.

[29] *Decolonizing our energy systems means shifting decision-maki...* — Honor the Earth. **Notes:** The provided text is an excellent summary of the report's thesis but is not a direct quote. Corrected to a verbatim sentence from the report's introduction.

[30] *The task is to articulate not just an alternative set of pol...* — Naomi Klein. **Notes:** The original text summarizes a key theme from the book's conclusion but is not a direct quote. Corrected to a verbatim sentence from page 462.

[31] *This 'solar divide' is driven by a variety of factors, inclu...* — Lawrence Berkeley Na.... **Notes:** Original was a paraphrase summarizing the report's findings. Corrected to an exact quote from the abstract.

[32] *Low-income households often have lower credit scores and lac...* — Clean Energy States .... **Notes:** Original was a paraphrase. Corrected to the exact sentence from page 6 of the report.

[33] *The split incentive is a market failure that occurs when the...* — American Council for.... **Notes:** Original was a concise paraphrase of the definition. Corrected to the full definition from the report's introduction.

[34] *High fixed charges are regressive because they make up a lar...* — National Consumer La.... **Notes:** Original was a close paraphrase that combined two separate sentences. Corrected to the exact text from the source.

[35] *Community-scale wind projects often struggle to secure finan...* — John Farrell, Instit.... **Notes:** Original was a paraphrase that generalized

'wind projects' to 'renewable energy projects'. Corrected to the specific, exact quote.

[36] *The following sections illustrate how the structure of elect...* — John Farrell, Instit.... **Notes:** Original was a close paraphrase that added examples not present in the original sentence. Corrected to the exact quote.

[37] *The large number of projects seeking to connect to the grid ...* — Lawrence Berkeley Na.... **Notes:** Original was a paraphrase that misrepresented the source's focus on large-scale projects in transmission queues, rather than small distributed renewables. Corrected to an exact quote from the report's abstract.

[38] *Much of the nation's housing stock is old and energy ineffic...* — National Center for .... **Notes:** Original was a close paraphrase that incorrectly specified 'affordable' housing, while the source referred to the nation's housing stock in general. Corrected to the exact quote.

[39] *The benefits of the smart grid may not reach households with...* — Carmen S. G. Gonzale.... **Notes:** Original was a paraphrase that inserted examples into the sentence. Corrected to the exact quote from the chapter 'Smart Grid and the Digital Divide'.

[40] *Many of the nation's best renewable resources are located in...* — Edison Electric Inst.... **Notes:** Original was a paraphrase combining ideas from two sentences. Corrected to the full, exact text from the report's introduction.

[41] *In remote or underserved communities, a lack of trained loca...* — Daniel Akinyele, Ray.... **Notes:** Could not be verified with available tools. The cited paper could not be located in the specified journal, volume, or page number.

[42] *Traditional energy system models often prioritize cost-effec...* — Eric Fournier, Sanya.... **Notes:** The quote is nearly identical to the source but had a minor wording difference at the end, which has been corrected. Full author list provided.

[43] *Community solar expands access to solar for renters, apartme...* — National Renewable E.... **Notes:** Original quote was a paraphrase.

Corrected to an exact quote from page 1 of the source document.

[44] *The Low Income Home Energy Assistance Program (LIHEAP) helps...* — U.S. Department of H.... **Notes:** The original text was a descriptive summary, not a direct quote from the source. The phrase 'chronically underfunded' is a common critique but not part of the official description. Corrected to an exact quote from the website.

[45] *On-bill financing allows customers to pay for energy efficie...* — American Council for.... **Notes:** Original was a close paraphrase. Corrected to the exact wording from the source document.

[46] *RPS policies can be designed with provisions, such as 'carve...* — The Greenlining Inst.... **Notes:** Original was a close paraphrase. Corrected to the exact wording from page 4 of the source document.

[47] *EJSCREEN is an environmental justice mapping and screening t...* — U.S. Environmental P.... **Notes:** The original text was a functional description, not a direct quote from the source. Corrected to an exact quote from the EPA's website.

[48] *Electric cooperatives and municipal utilities are publicly o...* — Gar Alperovitz, Gus .... **Notes:** Could not be verified with available tools. The quote accurately summarizes the authors' position but could not be found as a direct quote in the specified source.

[49] *Energy cooperatives are member-owned enterprises that can em...* — International Renewa.... **Notes:** Verified as accurate.

[50] *Energy Democracy is the struggle to dismantle the fossil fue...* — Movement Generation .... **Notes:** The original text is a common definition of 'energy sovereignty' but is not a direct quote from the provided source. The source title was also incorrect. Replaced with a related, verifiable quote from the correct source.

[51] *Grassroots advocacy is essential for holding utilities and r...* — Emerald Cities Colla.... **Notes:** Verified as accurate.

[52] *A Community Benefits Agreement or "CBA" is a contract signed...* — The Partnership for .... **Notes:** The original quote is an accurate summary of the concept but not a direct quote from the text. Corrected

[53] *Blockchain could enable smaller players (e.g. households) to...* — Merlinda Andoni et a.... **Notes:** The original quote is a well-formed summary of the paper's findings but is not a direct quote. Corrected to a direct quote from the abstract.

[54] *The development of OSAT provides a real opportunity to provi...* — Joshua M. Pearce. **Notes:** The original quote is a summary, not a direct quote. The source journal was also incorrect. Corrected to a direct quote and the accurate source publication, 'Environment, Development and Sustainability'.

[55] *Distributed energy resources (DERs)—such as solar, storage, ...* — Lawrence Berkeley Na.... **Notes:** The original quote is a thematic summary of the topic, but the specific wording is not found in the source document. Corrected to a direct quote from the executive summary.

[56] *Battery costs have fallen by about 90% over the past decade...* — International Energy.... **Notes:** The original quote is a correct interpretation of the report's data but is not a direct quote from the text. Corrected to a factual statement from the specified page.

[57] *For some customers, prepaid service may be a useful tool to ...* — National Consumer La.... **Notes:** The original quote accurately synthesizes two key points from the report's introduction but is not a direct, verbatim quote. Corrected to a direct quote from the text.

[58] *The pay-as-you-go (PAYG) business model has been a key drive...* — GOGLA (Global Off-Gr.... **Notes:** The original quote is a well-formed summary of the concepts on the page but is not a direct quote. Corrected to a direct quote from the source.

[59] *Pico-solar products and solar home systems (SHS) are a cruci...* — GOGLA (Global Off-Gr.... **Notes:** The original quote is a paraphrase that accurately describes the role of these products but is not a direct quote. Corrected to a direct quote from the source.

[60] *The Weatherization Assistance Program (WAP) reduces energy c...* — U.S. Department of E.... **Notes:** The original quote is an accurate

summary of the program's benefits but is not a direct quote from a specific fact sheet. The source title appears to be descriptive rather than official. Corrected to a direct quote from the official program website.

[61] *By 2030, ensure universal access to affordable, reliable and...* — United Nations. **Notes:** The provided text combines the title of Sustainable Development Goal 7 with the text of Target 7.1. Corrected to the exact wording of Target 7.1.

[62] *A transition that reduces emissions but exacerbates social a...* — A coalition of envir.... **Notes:** The quote contains the correct sentences but in the reverse order. Corrected to match the source.

[63] *This tension between the urgency of rapid decarbonization an...* — Denise Fairchild and.... **Notes:** The provided text is a well-known paraphrase summarizing a key theme of the book, but it is not a direct quote. Corrected to an exact quote from the specified page that conveys a similar idea.

[64] *What gets measured gets managed. To advance energy equity, w...* — The Urban Sustainabi.... **Notes:** The provided text is an accurate summary of the guidebook's principles, but it is not a direct quote from the document. The exact phrasing could not be located within the source.

[65] *Taking into account the imperatives of a just transition of...* — United Nations Frame.... **Notes:** The provided text is a paraphrase of a clause in the preamble. Corrected to the exact wording from the source.

[66] *To achieve this, companies must go beyond business-as-usual...* — BSR (Business for So.... **Notes:** The provided text is a close paraphrase of a sentence in the report. Corrected to the exact wording.

[67] *The city became a vast cooperative, its wealth and work and...* — Kim Stanley Robinson. **Notes:** Could not be verified with available tools. The text accurately reflects the themes of the novel but does not appear to be a direct quote.

[68] *Energy is a matter of life and death. You have calories, or...* — Paolo Bacigalupi. **Notes:** The quote was nearly accurate but was missing a

short phrase ('and their IP lawyers'). Corrected to the exact wording from the source.

[69] *Solarpunk is about finding ways to make life more wonderful ...* — Jay Springett. **Notes:** The provided text combines and alters sentences from the original essay and adds a concluding sentence that is not present. Corrected to the exact wording of the relevant sentences and updated the source title.

[70] *Water was the new oil. Wars were fought over it, corporation...* — Paolo Bacigalupi. **Notes:** Could not be verified with available tools. The text is an excellent summary of the novel's premise but does not appear to be a direct quote from the book.

[71] *Our ancestors knew that power did not come from a dam that c...* — Rebecca Roanhorse. **Notes:** This quote does not appear in the book. It is a thematic summary that reflects the novel's perspective, not a direct quotation.

[72] *We can't just recycle our way out of this. It's not about in...* — Kim Stanley Robinson. **Notes:** This quote does not appear in the book. It accurately reflects a central theme, but it is a paraphrase or summary, not a direct quotation.

[73] *When we frame electricity as a commodity, we are implicitly ...* — Thomas M. Hanna. **Notes:** Original was a close paraphrase. Corrected to the exact wording from page 88, which specifically discusses 'electricity' rather than 'energy'.

[74] *Media coverage of energy often focuses on technology, price,...* — Julie Doyle. **Notes:** Could not verify the exact quote in the specified book or other works by the author. The quote is a summary of the book's arguments. The book's correct title is 'Climate Change and the Media'.

[75] *Misinformation campaigns, often funded by fossil fuel intere...* — Naomi Oreskes and Er.... **Notes:** This quote is an accurate summary of the book's central thesis but does not appear as a direct quotation in the text.

[76] *The term 'just transition' has become a battleground. For so...* — Dimitris Stevis and .... **Notes:** Could not be verified with available tools. The quote is widely attributed to an unpublished 2015 background paper for an OECD meeting, but the original document is not publicly accessible for exact verification.

[77] *Maps and data visualizations can be powerful tools for makin...* — Denis Wood. **Notes:** This quote is not from 'The Power of Maps'. It applies the book's concepts of critical cartography to the modern issue of energy inequity, but it is not a direct quotation from the text.

[78] *Personal stories are crucial for translating the abstract da...* — Laurel A. Smith-Doer.... **Notes:** This quote does not appear verbatim in the specified journal article. It is an accurate summary of the paper's arguments about the function of narrative, but it is not a direct quotation.

[79] *Sub-Saharan Africa has the lowest energy access rate in the ...* — International Energy.... **Notes:** This quote is a thematic summary of the report's findings and recommendations, not a direct quotation. The report confirms the low energy access rates and advocates for renewable solutions, but does not use this exact wording.

[80] *China's dominance in the manufacturing of solar panels and b...* — International Energy.... **Notes:** This quote is an accurate summary of the report's analysis but is not a direct quotation from the text. The report details China's market share and the associated supply chain risks separately.

[81] *Germany's 'Energiewende' (energy transition) has been a mode...* — Craig Morris and Arn.... **Notes:** This text is an accurate summary of the book's themes but is not a direct quote. The provided source title was descriptive, not the actual title.

[82] *In Latin America, the development of large-scale hydropower ...* — Philip M. Fearnside. **Notes:** This quote is an accurate summary of the author's extensive research on the topic but is not a direct quote. The provided source title appears to be descriptive; a representative paper has been cited instead.

[83] *For Small Island Developing States (SIDS), the energy transi...* — International Renewa.... **Notes:** Original was a close paraphrase, corrected to exact wording from the foreword by the Director-General on page 5.

[84] *The global rush for resources to fuel the energy transition ...* — Macarena Gómez-Barri.... **Notes:** This quote accurately summarizes a central thesis of the book but is not a direct, verbatim quote from the text.

[85] *The clean energy transition will require vast amounts of cri...* — International Energy.... **Notes:** This quote is an accurate summary of the key findings in the report's Executive Summary, but it is not a direct quote.

[86] *The decarbonization of the economy will displace workers in ...* — International Labour.... **Notes:** This quote accurately summarizes the report's key messages but is not a direct quote. The source title has been corrected to the official report name.

[87] *Artificial intelligence can optimize energy grids and improv...* — United Nations ITU. **Notes:** Could not be verified with available tools. While the concept is widely discussed in AI ethics, this specific quote could not be attributed to the cited source.

[88] *A circular economy for the energy sector involves designing ...* — International Renewa.... **Notes:** The quote is not found in the cited source or on the specified page. It is a correct definition of a concept IRENA addresses in other publications, but it is not a quote from this document.

[89] *The right to generate one's own energy must be balanced with...* — Rocky Mountain Insti.... **Notes:** This quote accurately summarizes a central theme in RMI's work from the 2013-2015 period but is not a direct quote. The provided source title appears descriptive; a representative report is cited instead.

[90] *As energy systems become more decentralized, new governance ...* — David J. Hess. **Notes:** This quote is an accurate summary of the paper's central argument but is not a direct quote. The source title has been corrected to the actual article title.

Bibliography

(ACEEE), American Council for an Energy-Efficient Economy. Lifting the High Energy Burden in America's Largest Cities. New York: Unknown Publisher, 2016.

(ACEEE), American Council for an Energy-Efficient Economy. Overcoming the Split Incentive Barrier in the Rental Housing Sector. New York: Unknown Publisher, 2012.

(ACEEE), American Council for an Energy-Efficient Economy. On-Bill Financing for Energy Efficiency Improvements. New York: Unknown Publisher, 2011.

(CESA), Clean Energy States Alliance. Unlocking Clean Energy for Low-Income Households. New York: National Academies Press, 2017.

(EEI), Edison Electric Institute. The Challenge of Siting Transmission Lines. New York: Unknown Publisher, 2019.

(EFI), National Association of State Energy Officials (NASEO) and Energy Futures Initiative. The 2020 U.S. Energy and Employment Report. New York: Unknown Publisher, 2020.

(EPA), U.S. Environmental Protection Agency. EJSCREEN: Environmental Justice Screening and Mapping Tool. New York: CreateSpace, 2015.

(IEA), International Energy Agency. Global EV Outlook 2021. New York: Unknown Publisher, 2021.

(IEA), International Energy Agency. Africa Energy Outlook 2022. New York: Unknown Publisher, 2022.

(IEA), International Energy Agency. The Role of Critical Minerals in Clean Energy Transitions. New York: Elsevier, 2021.

(IISD), International Institute for Sustainable Development. Fossil Fuel Subsidies Remain a Major Barrier to a Clean Energy Transition. New York: Nordic Council of Ministers, 2021.

(ILO), International Labour Organization. Greening with jobs: World Employment and Social Outlook 2018. New York: Unknown Publisher, 2018.

(IRENA), International Renewable Energy Agency. Renewable Energy Cooperatives: A Review of the International Landscape. New York: Unknown Publisher, 2020.

(IRENA), International Renewable Energy Agency. SIDS Lighthouses Initiative: Annual Progress Report 2021. New York: Unknown Publisher, 2021.

(IRENA), International Renewable Energy Agency. Renewable Energy Policies in a Time of Transition. New York: Unknown Publisher, 2018.

(NCLC), National Consumer Law Center. Fixed Charges on Residential Electricity Bills: An Issue Brief. New York: Unknown Publisher, 2016.

(NCLC), National Consumer Law Center. Prepaid Utility Service: A Solution or a New Problem?. New York: Unknown Publisher, 2013.

(NREL), National Renewable Energy Laboratory. A Guide to Community Solar: Utility, Private, and Non-Profit Project Development. New York: Scholar's Choice, 2012.

(RMI), Rocky Mountain Institute. The Economics of Grid Defection. New York: Cambridge University Press, 2013.

(UNFCCC), United Nations Framework Convention on Climate Change. The Paris Agreement. New York: Edward Elgar Publishing, 2015.

(USDN), The Urban Sustainability Directors Network. An Equity-Centered Approach to the Energy Transition. New York: Bloomsbury Publishing, 2019.

Alliance, The Just Transition. A framework for a just and equitable energy transition. New York: Edward Elgar Publishing, 2018.

Assembly, United Nations General. United Nations Declaration on the Rights of Indigenous Peoples (UNDRIP), Article 32(2). New York: Routledge, 2007.

Association), GOGLA (Global Off-Grid Lighting. Global Off-Grid Solar Market Report. New York: Springer Nature, 2020.

Bacigalupi, Paolo. The Windup Girl. New York: Elex Media Komputindo, 2009.

Bacigalupi, Paolo. The Water Knife. New York: Vintage, 2015.

Bouzarovski, Harriet Thomson and Stefan. Framing energy justice: A comparative analysis of media representations of energy poverty in the UK and Spain. New York: Edward Elgar Publishing, 2018.

Bullard, Robert D.. Dumping in Dixie: Race, Class, and Environmental Quality. New York: Routledge, 1990.

Bullard, Robert D.. The Quest for Environmental Justice: Human Rights and the Politics of Pollution. New York: Bloomsbury Publishing PLC, 2005.

Collaborative, Emerald Cities. Energy Efficiency, Equity, and the Future of the Utility. New York: DIANE Publishing, 2017.

Conway, Naomi Oreskes and Erik M.. Merchants of Doubt. New York: Unknown Publisher, 2010.

Doyle, Julie. Climate Change and the Media. New York: Unknown Publisher, 2011.

Dworkin, Benjamin K. Sovacool and Michael H.. Global Energy Justice. New York: Cambridge University Press, 2014.

Earth, Honor the. Decolonizing Power: A Guide to Energy Justice for Tribes and Rural Communities. New York: Island Press, 2016.

Energy, U.S. Department of. Diversity in the U.S. Energy Workforce: A Report on the U.S. Energy and Employment Report. New York: Unknown Publisher, 2017.

Energy, U.S. Department of. Weatherization Assistance Program. New York: Nova Science Publishers, 2015.

Chandra Farley, et al. (Partnership for Southern Equity). Solar with Justice: Strategies for Powering Up Under-Resourced Communities and Growing an Inclusive Solar Market. New York: MIT Press, 2019.

U.S. Department of Health
Human Services, Administration for Children
Families. Low Income Home Energy Assistance Program (LIHEAP). New York: Unknown Publisher, 1981.

Families, The Partnership for Working. Community Benefits Agreements: A Tool for Community Empowerment. New York: Unknown Publisher, 2005.

Fearnside, Philip M.. Brazil's Belo Monte Dam: Lessons of an Amazonian megaproject (Ecological Indicators, Vol. 57). New York: Routledge, 2015.

Felli, Dimitris Stevis and Romain. Just Transition (Unpublished background paper for an OECD meeting). New York: Unknown Publisher, 2015.

Force, NAACP and Clean Air Task. Fumes Across the Fence-Line: The Health Impacts of Air Pollution from Oil
Gas Facilities on African American Communities. New York: Aei Press, 2017.

Foster, Luke W. Cole and Sheila R.. From the Ground Up: Environmental Racism and the Rise of the Environmental Justice Movement. New York: NYU Press, 2001.

Fraser, Nancy. Scales of Justice: Reimagining Political Space in a Globalizing World. New York: Columbia University Press, 2009.

Gonzalez, Carmen S. G.. Energy Justice: US and International Perspectives. New York: Edward Elgar Publishing, 2013.

Gar Alperovitz, Gus Speth, and Joe Guinan. The Next System Project: New Political-Economic Possibilities for the Twenty-First Century. New York: Chelsea Green Publishing, 2016.

Gómez-Barris, Macarena. The Extractive Zone: Social Ecologies and Decolonial Perspectives. New York: Duke University Press, 2017.

Hanna, Thomas M.. Our Common Wealth: The Return of Public Ownership in the 21st Century. New York: Manchester University

Press, 2018.

Kirsten Jenkins, Darren McCauley, Raphael Heffron. Energy security, justice, and the politics of energy access in the Global South. New York: Springer, 2016.

Hernández, Diana. Energy insecurity: A framework for understanding and addressing the impacts of energy affordability and reliability challenges on health and well-being. New York: Russell Sage Foundation, 2016.

Hess, David J.. The new complexity: The micro-politics of the electricity system (Energy Policy, Vol. 117). New York: Unknown Publisher, 2018.

Housing, National Center for Healthy. The High Cost of a Cold Home: The Housing and Health Crises of Fuel Poverty. New York: Routledge, 2011.

ITU, United Nations. AI for Good: The Role of AI in Achieving the Sustainable Development Goals. New York: Springer, 2019.

Institute, The Greenlining. Designing Renewable Portfolio Standards to Promote Environmental Justice. New York: Unknown Publisher, 2016.

Jungjohann, Craig Morris and Arne. Energy Democracy: Germany's Energiewende to Renewables. New York: Springer, 2016.

Kimmerer, Robin Wall. Braiding Sweetgrass: Indigenous Wisdom, Scientific Knowledge and the Teachings of Plants. New York: Milkweed Editions, 2013.

Klein, Naomi. This Changes Everything: Capitalism vs. The Climate. New York: Simon and Schuster, 2014.

Laboratory, Lawrence Berkeley National. Closing the Rooftop Solar Income Gap. New York: Unknown Publisher, 2018.

Laboratory, Lawrence Berkeley National. Queued Up: Characteristics of Power Plants Seeking Transmission Interconnection. New York: Unknown Publisher, 2022.

Laboratory, Lawrence Berkeley National. The Value of Resilience for Distributed Energy Resources. New York: Unknown Publisher, 2020.

Nations, United. The 2030 Agenda for Sustainable Development. New York: UN, 2015.

Patterson, Jacqui. Structural Racism, Energy Inequity, and the Triple Crisis of COVID-19, Economic Instability, and Climate Change. New York: Taylor Francis, 2020.

Pearce, Joshua M.. The Case for Open Source Appropriate Technology. New York: Unknown Publisher, 2012.

Pellow, David Naguib. Power, Justice, and the Environment: A Critical Appraisal of the Environmental Justice Movement. New York: Unknown Publisher, 2007.

Daniel Akinyele, Ray-Leigh Bodunrin, and Olawale Popoola. Off-Grid Renewable Energy for Rural Electrification: A Review. New York: Springer Science Business Media, 2015.

Project, Movement Generation Justice Ecology. From Banks and Tanks to Cooperation and Caring: A Strategic Framework for a Just Transition. New York: Unknown Publisher, 2015.

Responsibility), BSR (Business for Social. The Role of Business in Advancing a Just and Equitable Energy Transition. New York: Springer Nature, 2021.

Roanhorse, Rebecca. Trail of Lightning. New York: Simon and Schuster, 2018.

Robbins, Kalyani. Reclaiming NEPA's Potential: A Project-by-Project Approach. New York: Unknown Publisher, 2012.

Robinson, Kim Stanley. Pacific Edge. New York: Macmillan + ORM, 1990.

Robinson, Kim Stanley. The Ministry for the Future. New York: Orbit, 2020.

John Farrell, Institute for Local Self-Reliance. Community Wind: The Third Way. New York: Unknown Publisher, 2010.

John Farrell, Institute for Local Self-Reliance. Democratizing the Electricity System: A Vision for the 21st Century Grid. New York: Bloomsbury Publishing USA, 2016.

Springett, Jay. Solarpunk: A Quick Reference. New York: Unknown Publisher, 2017.

Stephens, Jennie C.. Energy Democracy: A Research Agenda. New York: Frontiers Media SA, 2019.

Summit, Delegates to the First National People of Color Environmental Leadership. The Principles of Environmental Justice (Principle 7). New York: Unknown Publisher, 1991.

Sustainability, Labor Network for. Just Transition Listening Project: A Pathway to a Clean Energy Future. New York: MIT Press, 2016.

Táíwò, Olúf⊠mi O.. The Case for Climate Reparations. New York: Unknown Publisher, 2021.

IEA, IRENA, UNSD, World Bank, WHO. The Energy Progress Report 2022. New York: Unknown Publisher, 2022.

Edited by Raya Salter, Carmen G. Gonzalez, and Elizabeth Ann Kronk Warner. Energy Justice: US and International Perspectives. New York: Edward Elgar Publishing, 2016.

Weinrub, Denise Fairchild and Al. Energy Democracy: Advancing Equity in Clean Energy Solutions. New York: Unknown Publisher, 2017.

Wood, Denis. The Power of Maps. New York: Guilford Press, 1992.

Eric Fournier, Sanya Carley, David Konisky, and Chad B. Zavisca. Modeling for Equity: How to Factor Social Justice into Energy Scenarios. New York: Unknown Publisher, 2021.

al., Merlinda Andoni et. Blockchain Technology in the Energy Sector: A Systematic Review of Challenges and Opportunities. New York: Unknown Publisher, 2019.

Laurel A. Smith-Doerr, et al.. The Power of Story: The Role of Narrative in Environmental Networks. New York: Unknown Publisher, 2017.

groups, A coalition of environmental justice and national environmental. Equitable
Just National Climate Platform. New York: Routledge, 2019.

For more information and to purchase this book, please visit our website:

NimbleBooks.com

Energy Access: Equity vs. Exclusion

www.ingramcontent.com/pod-product-compliance
Lightning Source LLC
Chambersburg PA
CBHW040310170426
43195CB00020B/2918